SYMBIOSIS | ˌsɪmbɪˈəʊsɪs, -bʌɪ-|
noun (pl. symbioses |-siːz|)

> interaction between two or more different species living in
> close physical association, typically to the advantage of each.

Over the past 4 billion years, microbes have shaped our Earth into the biosphere we know and love – rich in biological and geological diversity.

Through a range of symbioses (some brief, some lifelong), microbes have collaborated with all types of life on Earth to create new, emergent forms, including human beings. While some symbioses cause harm, most bring benefits to all involved.

The idea that life evolves through competition is only part of the story. Life is just as much about working together.

Created by Briony Barr & Dr. Gregory Crocetti
Written by Ailsa Wild
Illustrated by Aviva Reed
Art Direction & Additional Illustrations by Briony Barr
Storyboarding by Briony Barr & Dr. Gregory Crocetti
Scientific Writing by Dr. Gregory Crocetti & Briony Barr
Scientific Consulting from Dr. Karin Pritsch, Assoc. Prof. Jeff Powell,
Dr. Viviane Radl, Dr. Thorsten Grams, Dr. Karen Barry, Prof. Marc-André Selosse,
Dr. Paul Dennis, Dr. Chanyarat Paungfoo-Lonhienne,
Dr. Sapphire McMullan-Fisher, Prof. Linda Blackall & Dr. Pauline Mele
Book Design by Jaye Carcary
Artwork documentation by Theresa Harrison

Special thanks to Cheryl Power, Dianne Lightfoot, Eastmint Studios, Peter Barr,
James Murdoch, Jill Farrar, Monique Henry, Beth Askham, Kaela Drew,
Jono La Nauze, Patrick Belford, Jack Wild, Kaisha Reed and the year 3–6 students
at Thornbury Primary School and Lorne P-12 College.

The creation of this book was supported by the Australian Society for Microbiology
and by the Victorian Government through Creative Victoria.

This book is dedicated to farmers all over the world.

Unless otherwise attributed, the images on pages 36, 37, 38, 44 and 45 are sourced from Wikimedia Commons, Pexels, Pxhere, Science Photo Library and Adobe Stock Images.

A catalogue record for this book is available from the National Library of Australia.

ISBN: 9781486313310 (hbk.)
ISBN: 9781486313327 (epdf)

Published in collaboration with Scale Free Network by
CSIRO Publishing
Private Bag 10
Clayton South VIC 3169 Australia
Telephone: +61 3 9545 8400
Email: publishing.sales@csiro.au
Website: www.publish.csiro.au

Printed in China by Leo Paper Products Ltd

The views expressed in this publication are those of the author and do not necessarily represent those of, and should not be attributed to, the publisher or CSIRO.

How to pronounce the names of the characters in the story:

ACTINS	ak-TINS
AZOES	AY-zohs
BROMA	BROH-mah
GLOMUS	GLOH-muss
MONAS	MOH-nuhs
TILIS	TIL-lees

This story is about microscopic fungi and bacteria in the soil. A teaspoon of healthy soil contains billions of bacteria and hundreds of metres of fungal hyphae (threads).

THE FOREST
IN THE TREE
HOW FUNGI SHAPE THE EARTH

Written by Ailsa Wild
Illustrated by Aviva Reed
Created by Briony Barr & Dr. Gregory Crocetti

PART ONE:
A SPORE UNDERGROUND

Water wakes me. I'm hungry.
I'm a tiny fungal spore and I'm carrying fats
and sugars. These are my survival rations…
but they'll only last a few days.
I need to find more food.

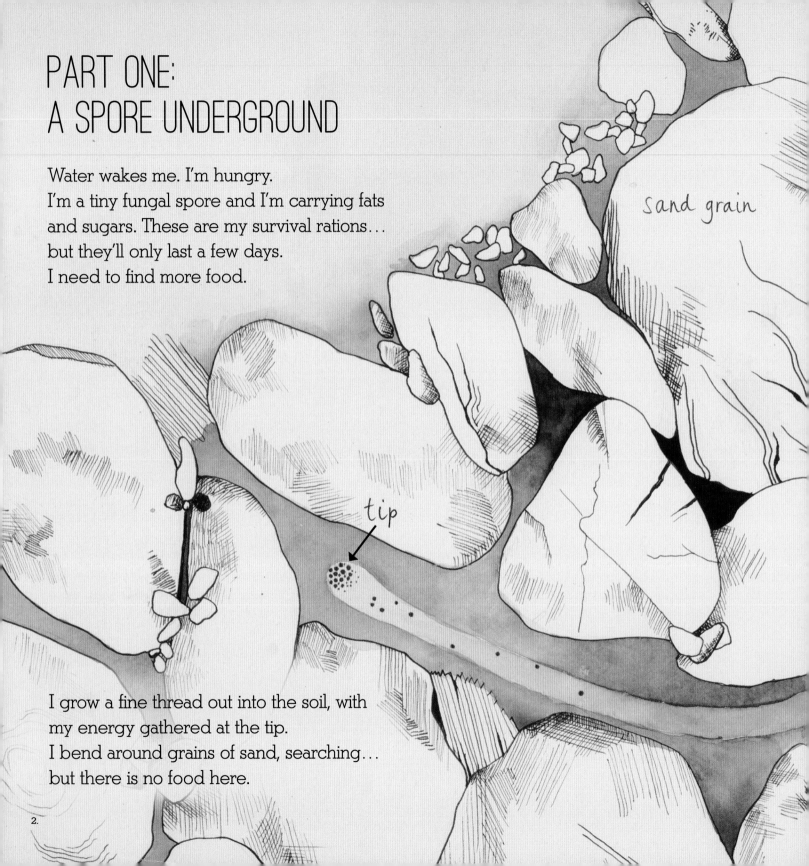

sand grain

tip

I grow a fine thread out into the soil, with
my energy gathered at the tip.
I bend around grains of sand, searching…
but there is no food here.

2.

I can't give up.
I grow another thread in a new direction.

The soil crackles with crystals covered in busy Tili bacteria. The Tilis are tiny and strong, using water droplets to release phosphorus from the soil.

"Greetings!" I call.
"Hello!" a few reply.
"Can you help me? Have you seen any plant roots?" I ask.
"No," they say, "not around here."

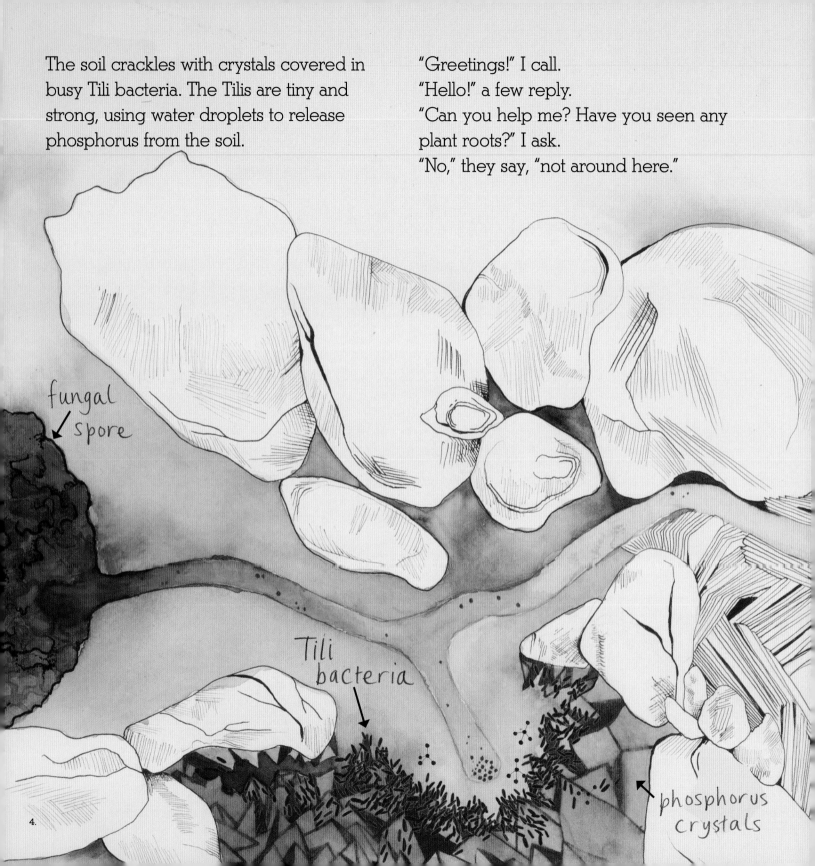

fungal spore

Tili bacteria

phosphorus crystals

I try a new way. I dig through clay.
I break rock.
But I'm running out of supplies.
If I don't find a plant soon I will starve.

message
molecule

And then, there's something…
a molecule…?
A message molecule!

clay

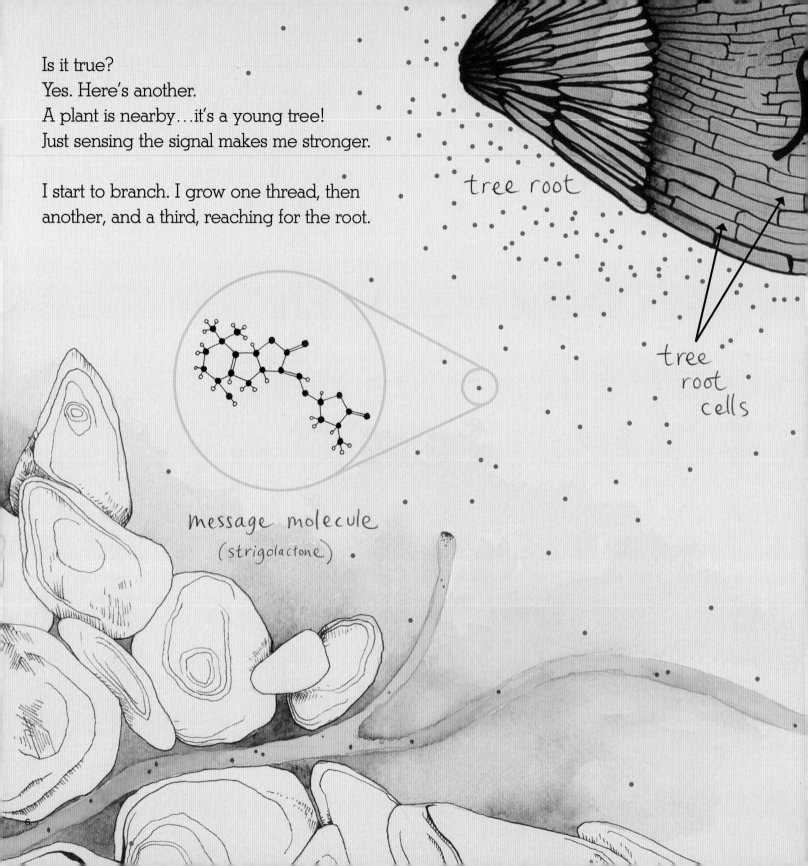

Is it true?
Yes. Here's another.
A plant is nearby…it's a young tree!
Just sensing the signal makes me stronger.

I start to branch. I grow one thread, then
another, and a third, reaching for the root.

tree root

tree
root
cells

message molecule
(strigolactone)

6.

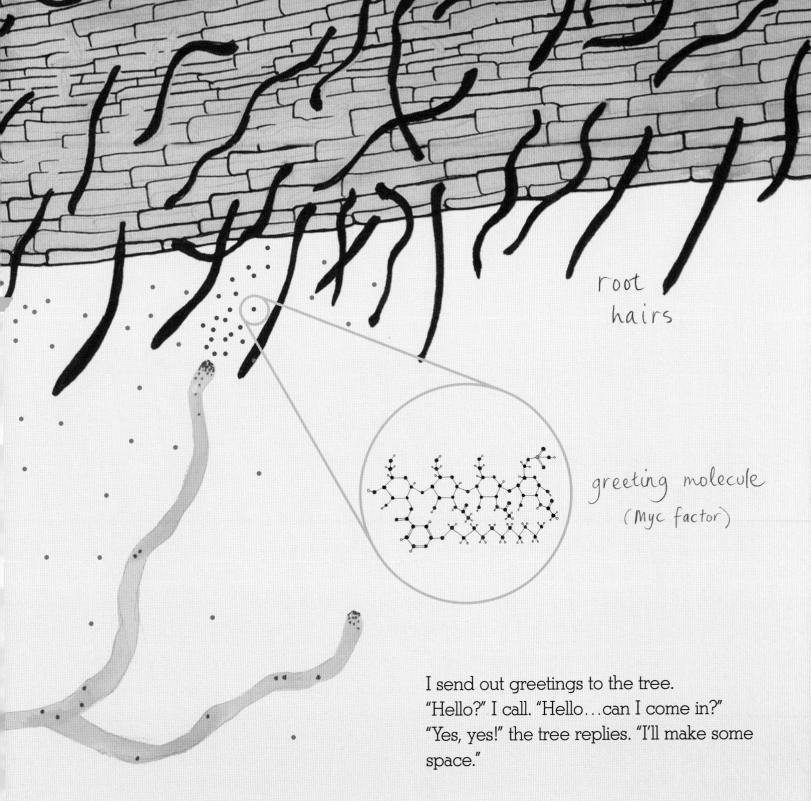

root hairs

greeting molecule
(Myc factor)

I send out greetings to the tree.
"Hello?" I call. "Hello…can I come in?"
"Yes, yes!" the tree replies. "I'll make some space."

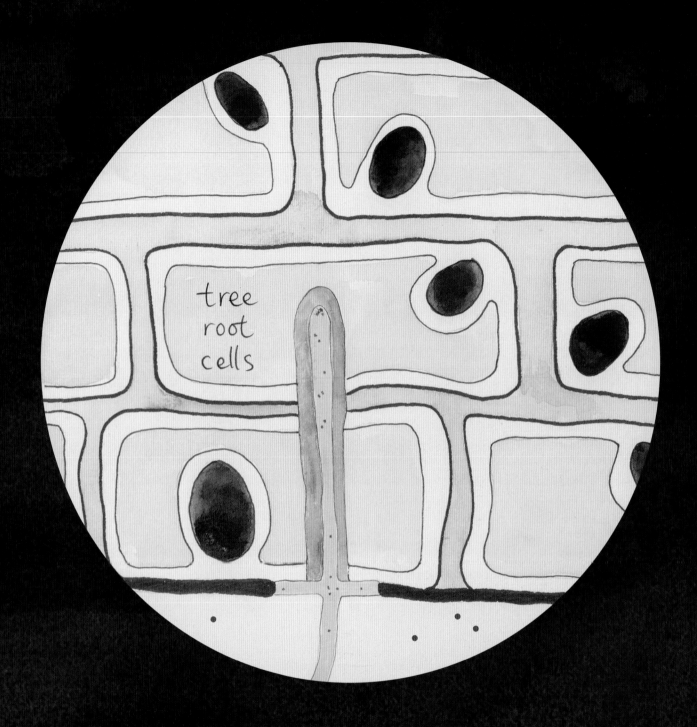

My first thread connects. I spread out, building a little platform on the root.

I find a tunnel and follow it into a root cell – a little place just perfect for me.

Here, I branch and branch again until I'm like a tiny tree inside the tree.

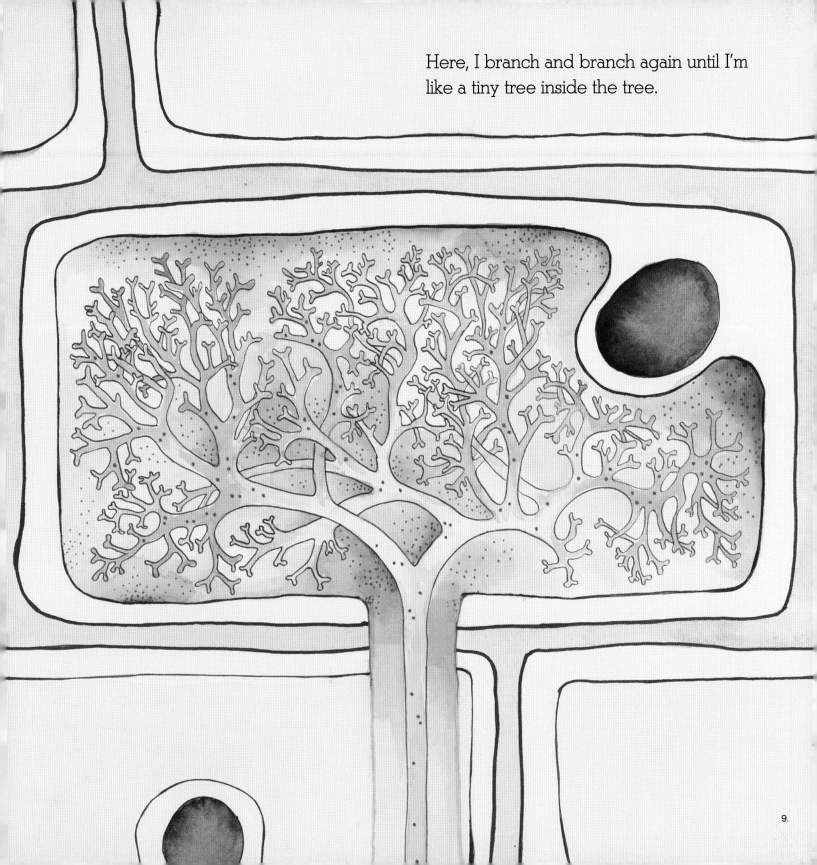

Broma

25 cm

sugar
(glucose)

spore

hyphae

plant roots

Soil

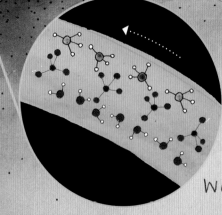

nutrients
water

The tree talks to me. She's alone and very
young, clinging to poor, dry soil.
"I'm Broma," she says.
She sends sugar, which gives me strength.

"I'll help you, but I need your help too,"
she says. "I need nutrients and water.
Otherwise we'll both die."
"I'll do my best," I promise.

I send my threads into the soil, reaching
into places that are too small for Broma's
roots. I find nutrients and water and carry
them, molecule by molecule, to feed her.

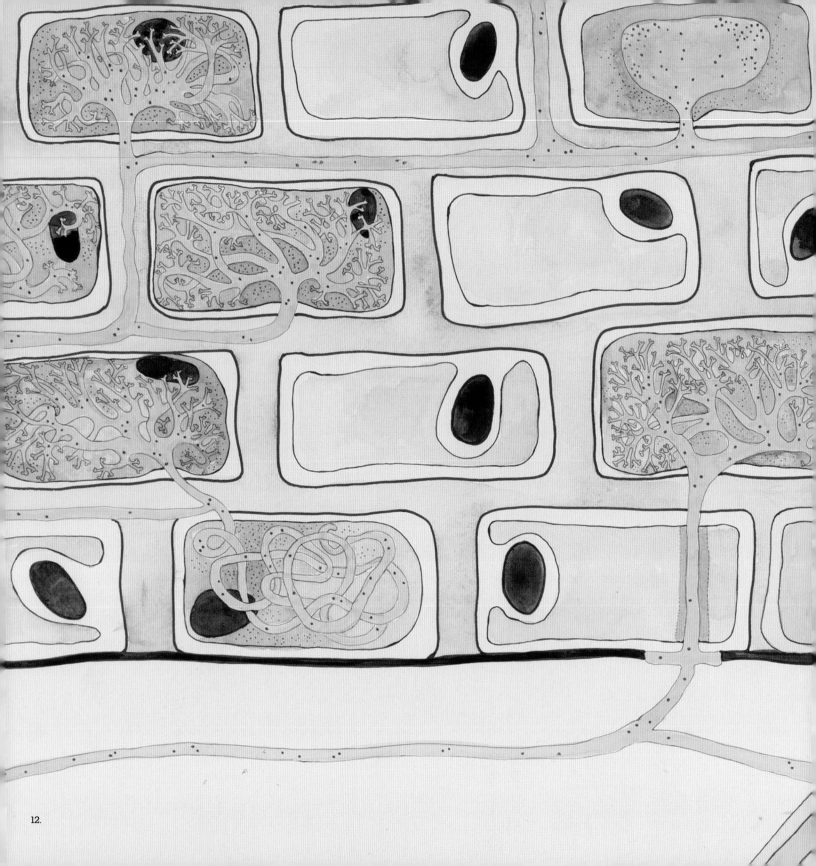

12.

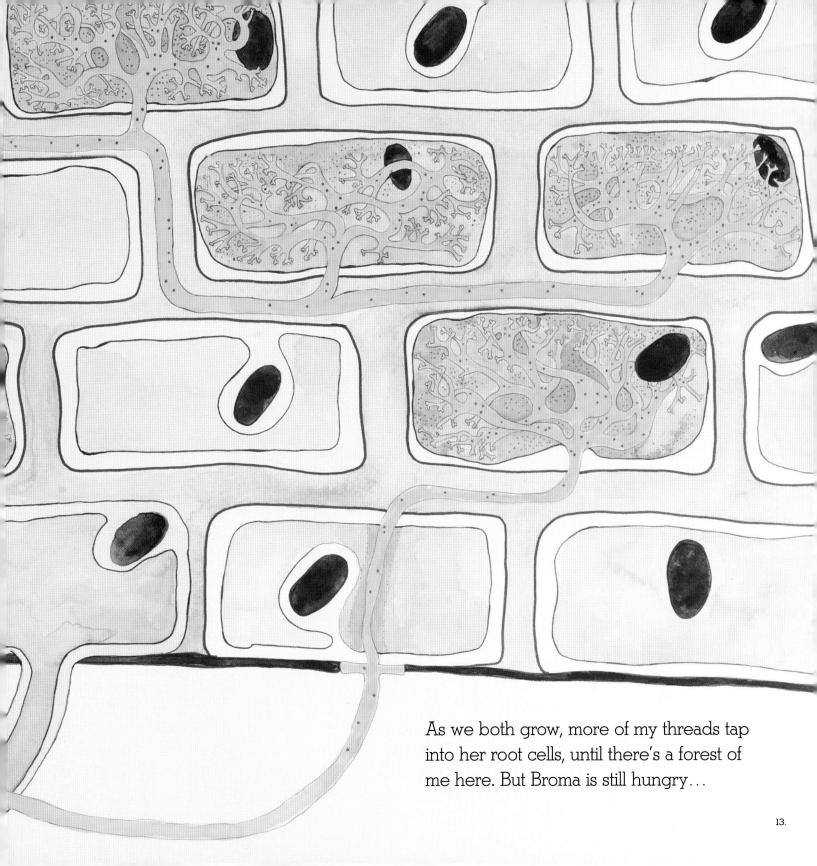

As we both grow, more of my threads tap into her root cells, until there's a forest of me here. But Broma is still hungry...

13.

I seek further and further through the soil,
until one of my tips senses something.
"Greetings!" it calls.
It's another fungus, like me.
"Greetings," I respond, curiously.

We touch. It's like destiny. It's like melting.
Our skins soften, and we swap parts of
ourselves.
Now I am the fungus and she is me.
"We are Glomus," she says.
"We are Glomus," I agree.

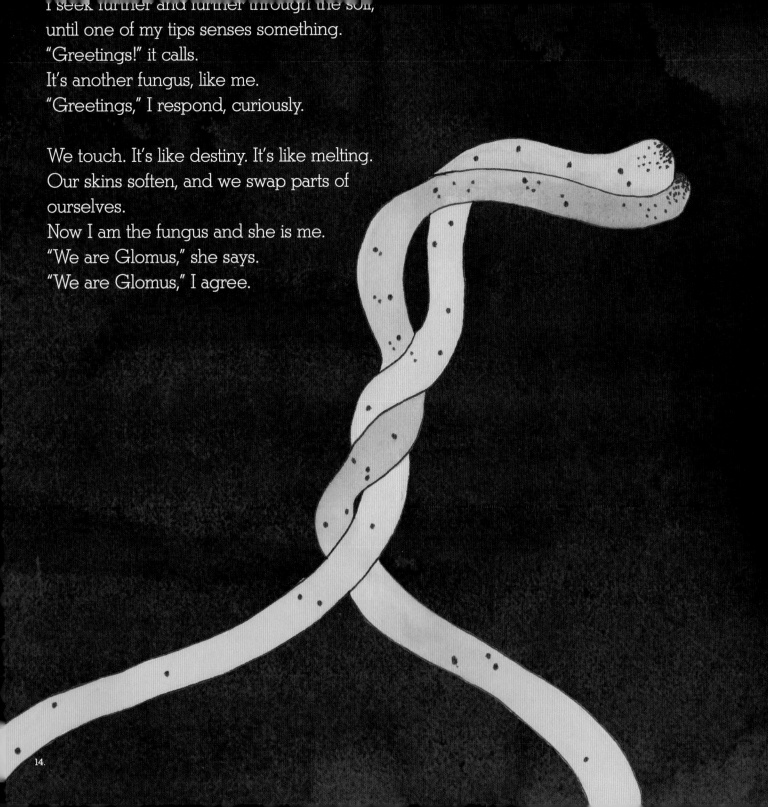

Broma

Glomus

15.

Broma

PART TWO:
WE, GLOMUS

Broma is older now. Her sensitive roots
forage in the soil on the edge of the forest.
She grows seeds and drops them each
year.

We, Glomus, have our threads in many
places at once – linked to hundreds of
other trees and smaller plants. We carry
water and nutrients from root to root,
helping plants to share resources.

We are part of an enormous forest web.

cacao
tree
(Theobroma
cacao)

16.

Glomus

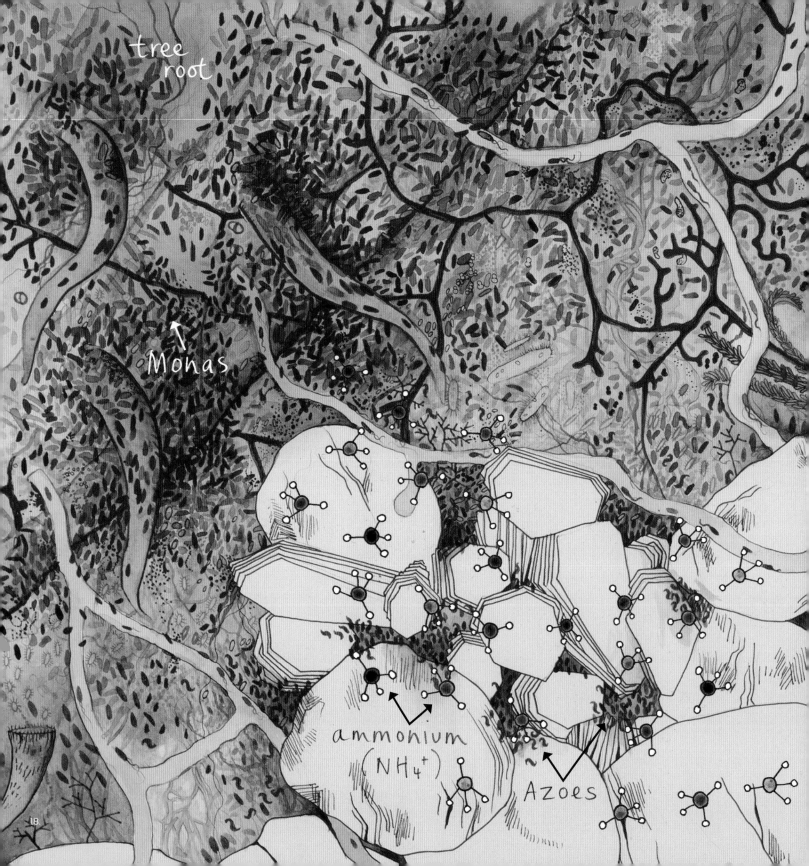

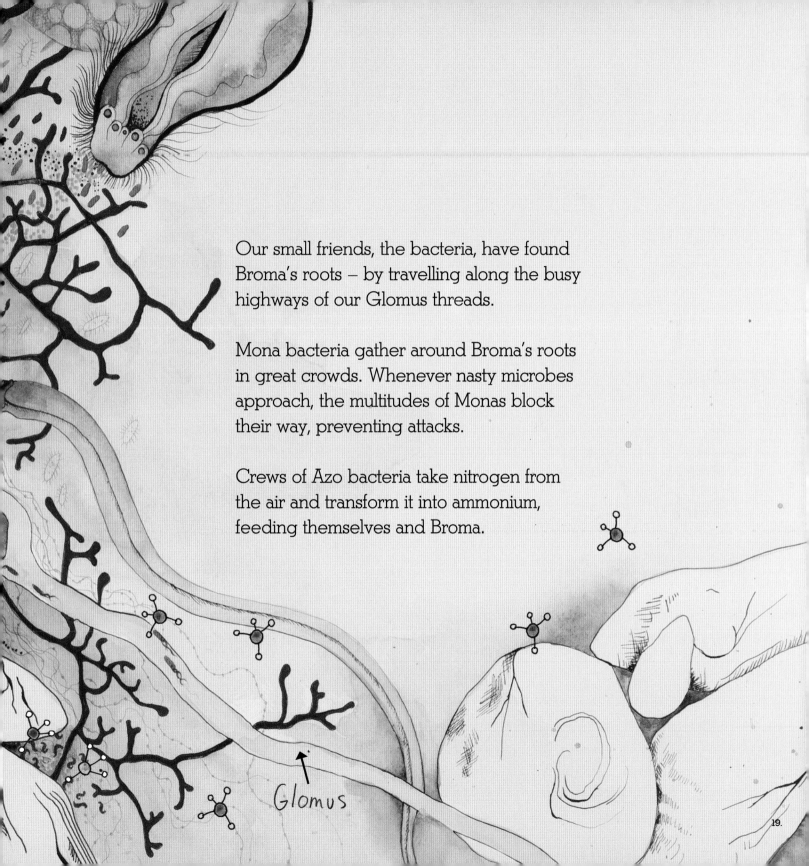

Our small friends, the bacteria, have found Broma's roots – by travelling along the busy highways of our Glomus threads.

Mona bacteria gather around Broma's roots in great crowds. Whenever nasty microbes approach, the multitudes of Monas block their way, preventing attacks.

Crews of Azo bacteria take nitrogen from the air and transform it into ammonium, feeding themselves and Broma.

Glomus

20.

Families of bacteria and fungi break down old leaves and bark and build lovely humus structures and tiny air pockets in the soil.

We, Glomus, curl and grow happily here, where the Actin bacteria make the darkness smell delicious. We work with the friendly bacteria, making the soil sticky and good to eat.

Together we shape the earth.

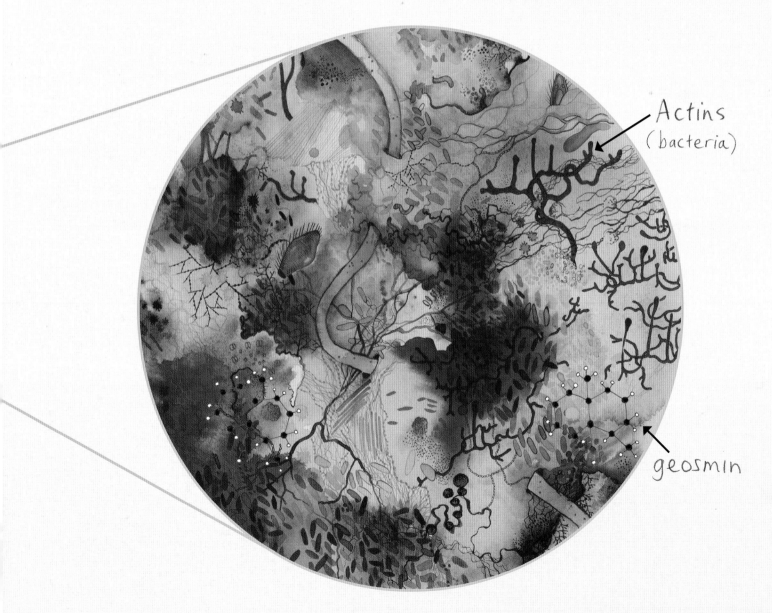

Actins
(bacteria)

geosmin

30 cm

One day our exploring Glomus threads
find a cluster of hungry baby trees,
growing far from the protective shade
of the forest. We tell Broma and she
thrills… it's her own children!

We send the children water and sugars,
humming through the web.
"Phosphorus!" they cry. "We need
phosphorus!"

We send out a thread to our old friends, the Tili bacteria. They are still dissolving crystals and pulling out phosphorus.
"We bring word from nearby trees," we say. "They can give you sugar, if you will share in return."

The delighted Tilis give us phosphorus for Broma's babies and they tumble over themselves to feast on our sugar.

All is well.

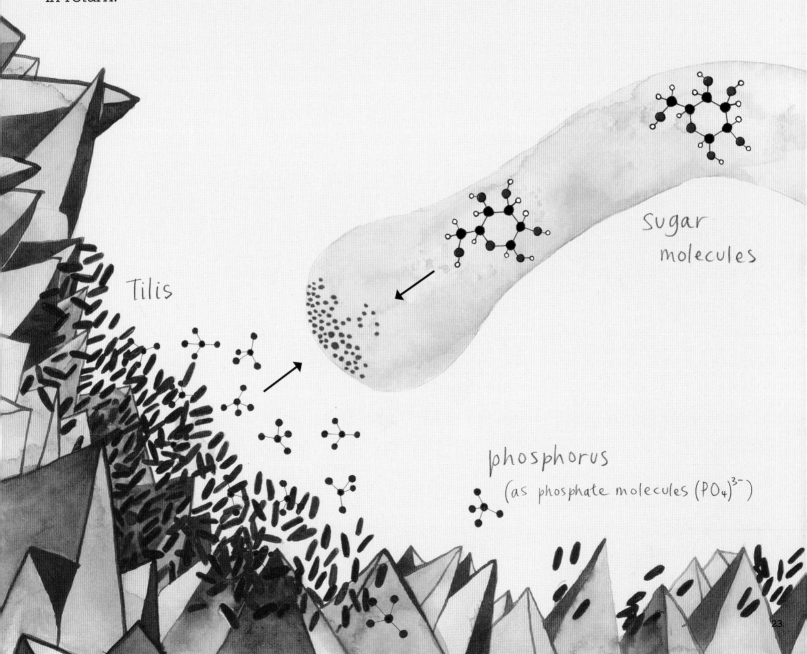

Tilis

sugar molecules

phosphorus
(as phosphate molecules $(PO_4)^{3-}$)

But that summer, the rains don't come.
For days we, Glomus, feel the heat
pounding down through the soil above.
The trees become thirsty.
"Water!" their roots demand. "Bring water!"

We, Glomus, take the moisture stored in
the soil along our length and give it to the
trees. We lessen their thirst, keeping them
alive. But they need more water. Weeks
pass. We're starting to feel desperate.

water
molecules

It's even worse for the baby trees who don't have enough small friends gathered around their roots to help.

The baby trees start to scream.

The soil here is still dry and hard and their roots are clumped so close together.

Our threads hunt desperately through the soil. Each time we find a tiny oasis of moisture, built by microbes, we share it with the trees.

But many microbes have stopped their building – things are going badly for the bacteria.

The Tilis need water for their mining, and the phosphorus supply grinds to a halt. Billions of Actins stop their tasks and become still. The Azoes all shut down into a waiting sleep. And the Monas are dying by the millions.

Around us, vibrations of screaming thirst beat through the tree roots.

Something is terribly wrong with one of the baby trees. There is pain and then… emptiness. Broma senses the loss of one of her babies.

We, Glomus, hold her. With the whole forest web, we hold her in a vast, tangled embrace.

We are together…and we are facing death.

Finally the soil begins to cool…can it be rain?

Yes! Trickles of water start to seep down through the darkness around us.

We feel tree roots take the rain, gratefully quenching their thirst.
The bacteria begin to move.

We reach a thread towards the baby tree.
We were able to keep so many trees alive.
But not this one. In the hot, dry wind it has fallen to the ground.

Billions of tiny friends gather along it, turning it into rich, sticky soil.

Broma's surviving children will have more food and space to grow.

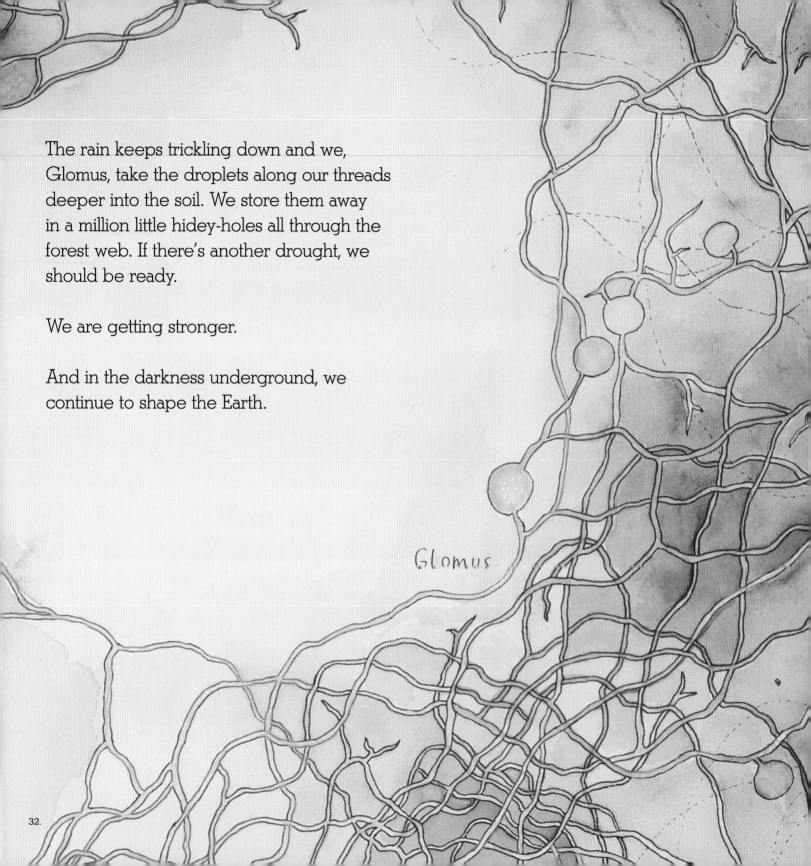

The rain keeps trickling down and we,
Glomus, take the droplets along our threads
deeper into the soil. We store them away
in a million little hidey-holes all through the
forest web. If there's another drought, we
should be ready.

We are getting stronger.

And in the darkness underground, we
continue to shape the Earth.

Glomus

THE SCIENCE BEHIND THE STORY

A SIMPLE GUIDE TO THE SYMBIOSIS IN THE STORY

Plants *TRANSFORM* energy from the Sun

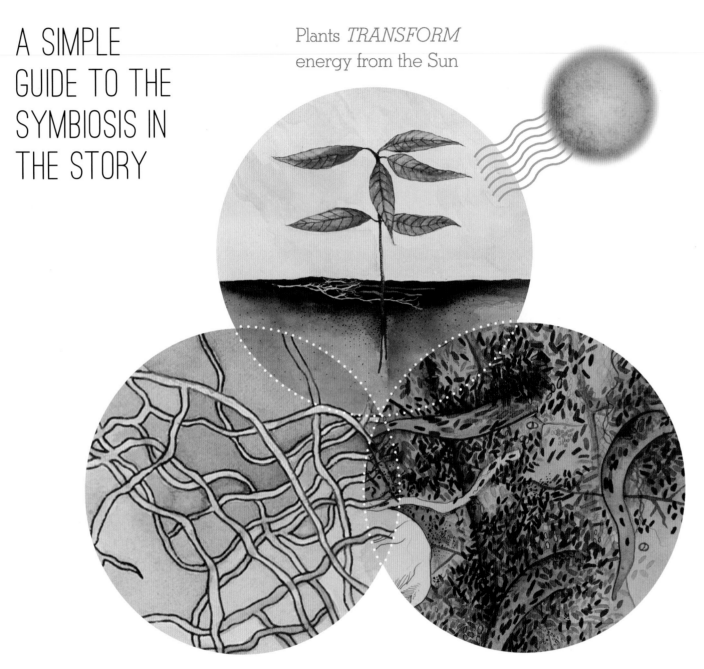

Mycorrhizal fungi *CONNECT + SHARE* energy, nutrients and water

Bacteria *RECYCLE* the nutrients

UNDERGROUND RELATIONSHIPS

All creatures require energy, nutrients and water in order to survive. To meet these needs, plants, fungi and soil microbes have worked together in symbiotic partnerships for over 400 million years.

PLANTS (BROMA)

Plants can transform carbon dioxide and water into sugars (and fats) using energy from sunlight. However, plants need help to gather water and nutrients (such as nitrogen and phosphorus) from the soil, so they exchange some of their energy-rich sugars for water and nutrients from bacteria and mycorrhizal fungi.

MYCORRHIZAL FUNGI (GLOMUS)

Mycorrhizal fungi grow kilometres of threads (called hyphae) from the plant roots out into the soil. These tiny threads are excellent at finding, absorbing and transporting hidden pockets of water and nutrients back to their plant partners in exchange for energy-rich sugars (and fats) – much like an extension of the plant's root system. These fungal networks also connect and share some of the plant's energy with other microbes, such as bacteria, in exchange for mineral nutrients from the soil.

BACTERIA (ACTINS/AZOES/MONAS/TILIS)

Bacteria, fungi and other soil microbes break down (decompose) minerals and organic matter (such as dead insects and fallen leaves) into basic building blocks, releasing valuable nutrients (such as nitrogen and phosphorus) back into the soil. Plants and mycorrhizal fungi help stimulate and power this important recycling work by sharing energy-rich sugars.

PLANTS: SOLAR ENERGY FOR THE SOIL

The fossil record suggests that plants first emerged from the sea as single-celled green algae about 500 million years ago. To survive in this new, dry landscape, the algae formed symbiotic partnerships with fungi and bacteria.

Since then, plants, fungi and bacteria have continued to co-evolve a range of different relationships, to colonise almost all land surfaces of the Earth – from forests and grasslands to gardens and city parks.

ENERGY FROM THE SUN

Plants transform energy from sunlight into chemical energy through a process called **photosynthesis**.

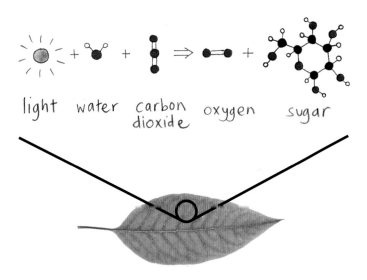

light   water   carbon dioxide   oxygen   sugar

For this reason, plants are considered the energy powerhouse driving most of the activities within the soil ecosystem.

FOOD FROM THE SOIL

The most important mineral nutrients required by plants are phosphorus (P) and nitrogen (N). Plants also require a range of other nutrients in lesser amounts, such as potassium (K), sulfur (S), calcium (Ca), magnesium (Mg) and zinc (Zn).

"...vibrations of screaming thirst beat through the tree roots." (pg27)

A growing body of scientific research suggests that many plants can both produce and react to sound waves. Under conditions of water stress (drought), some plants emit ultrasonic vibrations, described as "screaming" in this story.

"We send the children water and sugars, humming through the web." (pg22)

While most plants and larger trees appear aboveground as individuals competing for space, light and nutrients, they are often connected with each other below the ground, cooperating to share energy, water and nutrients through their networks of mycorrhizal fungi (the 'wood wide web') and sometimes directly between neighbouring roots. By working together, plants create more stable and resilient communities, such as forests and grasslands.

THEOBROMA CACAO: THE CHOCOLATE TREE

FOOD OF THE GODS

Theobroma cacao is a small tree, native to the tropical rainforests of the Amazon – the setting for this story. The genus name *Theobroma* derives from the Greek words *theos* (god) and *broma* (food). The species name *cacao* derives from the indigenous Mayan word *kakaw*.

Cacao trees naturally grow under the canopy of tropical rainforests. In this story, we meet Broma, a tree that has started to grow on the very edge of the forest, just beyond the protective shade and humidity of the nearby trees. This challenging situation makes Broma's partnership with her mycorrhizal fungi partner (*Glomus*) even more important.

Cacao flowers can produce fruits called cacao pods, containing 20–60 seeds (also called beans). Although traditionally prepared as a beverage, the beans from the cacao tree are now mostly used to make chocolate. Today, *Theobroma cacao* trees are grown across many equatorial regions around the globe, with nearly half of all cacao beans produced in the West African countries of Ivory Coast and Ghana. The future global production of cacao (for chocolate) is under threat from the spread of pests and diseases and a predicted increase in droughts due to climate change.

Where does Broma fit in the scientific classification of all life?

Domain:	*Eukarya*
Kingdom:	*Plantae (Archaeplastida)*
Clade:	*Angiosperms*
Clade:	*Eudicots*
Clade:	*Rosids*
Order:	*Malvales*
Family:	*Malvaceae*
Genus:	*Theobroma*
Species:	*cacao*

FANTASTIC FUNGI

Fungi play a central role in the recycling of carbon and other nutrients through ecosystems across the planet. Fungi are best known by the mushrooms that briefly appear from soil after rain. However, most fungi spend their entire lives underground, hidden from our sight.

Soil fungi get their energy in three main ways:

- **Saprotrophic (decomposer) fungi** work with bacteria to harness energy from breaking down dead organic matter.
- **Parasitic & Predatory fungi** steal energy by attacking living creatures, including many types of plants and animals.
- **Mycorrhizal fungi** (like *Glomus*) gain energy through their symbiotic partnerships with plant roots.

"I'm a tiny fungal spore and I'm carrying fats and sugars. These are my survival rations… but they'll only last a few days." (pg2)

A spore can be understood as the beginning and end of a fungal life cycle. Fungal spores are microscopic structures that fungi use to reproduce, similar to seeds. Spores can lay dormant (with their metabolism slowed) for long periods until woken by plant signals or water. Fungal spores vary in size and weight: some are light enough to be carried by wind and rain; but big, heavy spores (like *Glomus*) grow by spreading hyphal threads into the soil to find plant root partners.

"I grow a fine thread out into the soil, with my energy gathered at the tip." (pg2)

Most fungi (including *Glomus*) grow and travel as hyphae – long, thin, thread-like, tubular chains of cells containing many nuclei. Fungal hyphae can extend through the tiny pockets and pathways within soil, creating kilometres of branching networks called a mycelium. The direction and rate of growth of each hypha are coordinated by tiny structures at their tip, called the *Spitzenkörper* (German for 'pointed body').

MYCORRHIZAL FUNGI

The term mycorrhiza comes from the Greek words *mykes* (fungus) and *rhiza* (root) and is used to describe the mutually beneficial symbiotic partnership between a fungus and a plant. The hyphae of mycorrhizal fungi connect to plant roots in two main ways: forming a network around plant roots (called Ectomycorrhizae) or growing inside plant root cells (called Endomycorrhizae), like *Glomus*.

All mycorrhizal fungi rely on plant roots to provide them with energy-rich carbon, and in return supply plants with water and nutrients (particularly phosphorus) from the soil. This ancient and highly evolved relationship relies on the ability of both plants and fungi to recognise and reward beneficial symbiotic partners (called a positive feedback loop).

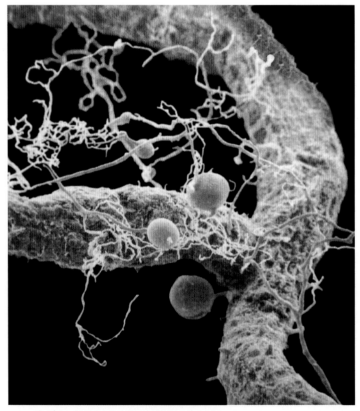

Scanning electron microscope image of mycorrhizal hyphae and spores around plant root.

WHEN GLOMUS MET BROMA

Glomus fungi belong to the broad group (phylum) *Glomeromycota*, one of the oldest lineages of fungi. These mycorrhizal fungi form a symbiosis with more than two-thirds of all known species of land plants, making it one of the most important symbiotic partnerships on Earth.

"And then, there's something…a molecule…? A message molecule! (pg5) "Just sensing the signal makes me stronger." (pg6)

Mycorrhizal fungi can detect special chemical signals (message molecules called strigolactones) released by plant roots. These signals attract the mycorrhizal fungi towards the plant and stimulate the growth and branching of fungal hyphae during the early germination phase.

"Here, I branch and branch again until I'm like a tiny tree inside the tree." (pg9)

Once inside the cells of a plant root, the hyphae of *Glomus* fungi form tiny tree-like structures where the fungus and the plant exchange nutrients. These structures are called arbuscules (from the Latin *arbusculum* for 'little tree') and give this group of fungi the name 'arbuscular mycorrhizal fungi'.

Image of arbuscule courtesy of Prof. Daniel Wipf, Université de Bourgogne, INRA Dijon, France.

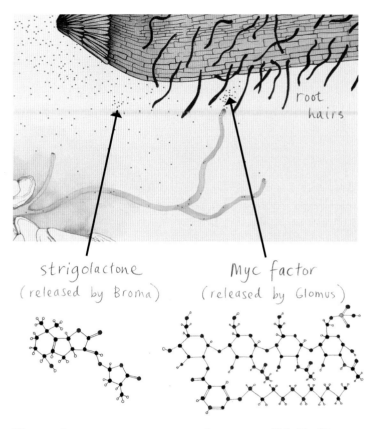

root hairs

strigolactone
(released by Broma)

Myc factor
(released by Glomus)

"I send out greetings to the tree. 'Hello?' I call. 'Hello…can I come in?' 'Yes, yes!' the tree replies. 'I'll make some space.'" (pg7)

In response to the signals from the plant, mycorrhizal fungi secrete their own chemical signals (called lipochitooligosaccharides or 'Myc factors'). If the plant recognises the signals as friendly, it will create a passage allowing the fungi to enter the root.

Where does Glomus fit in the scientific classification of all life?

Domain: *Eukarya*
Kingdom: *Mycota* (Fungi)
Phylum: *Glomeromycota*
Class: *Glomeromycetes*
Order: *Glomerales*
Family: *Glomeraceae*
Genus: *Glomus*

SMALL FRIENDS WITH SUPERPOWERS

Soil is home to an enormous variety of microbes such as viruses, archaea, fungi, protozoa and bacteria. Scientists estimate that a handful of healthy soil contains more bacteria than humans on Earth! Bacteria have the power to decompose, digest and dissolve organic and mineral matter back into basic building blocks that can be used by other organisms to grow. This story includes four examples of bacteria that play important roles in the soil ecosystem.

TILIS: THE ROCK SPLITTERS

"The Tilis are tiny and strong, using water droplets to release phosphorus from the soil." (pg4)

Phosphorus (P) is a key nutrient for growing plants. In soil, phosphorus atoms are usually attached in a group with four oxygen (O) atoms as phosphate molecules ($PO_4)^{3-}$. Phosphate binds strongly to surrounding molecules, making it difficult for most plants and fungi to access. However, bacteria, such as **Bacillus subtilis** (the 'Tilis'), make enzymes (active proteins) or acids to release phosphate back into a form that fungi and plants can absorb.

AZOES: THE NITROGEN FIXERS

"Crews of Azo bacteria take nitrogen from the air and transform it into ammonium, feeding themselves and Broma." (pg19)

Proteins play essential roles inside the cells of all life forms, and you can't build proteins without nitrogen. Even though nitrogen gas (N_2) makes up 78% of our atmosphere, plants cannot use it in this form. Some bacteria and archaea (called diazotrophs) have the special ability to convert atmospheric nitrogen (N_2) into a usable form such as ammonium (NH_4^+) through a process called nitrogen fixation. Some nitrogen-fixing bacteria live in a symbiotic partnership inside the roots of legume plants, while others, such as **Azospirillum** (the 'Azoes'), live freely in the soil. Once ammonium is created, it can be absorbed and used by plants and mycorrhizal fungi.

Image of soil microbes courtesy of the Soil and Terrestrial Environmental Physics group of Dani Or at ETH Zurich (www.step.ethz.ch). SEM image acquisition and coloring by Anne Greet Bittermann, ETH Zurich – ScopeM.

MONAS: THE BODYGUARDS

"Whenever nasty microbes approach, the multitudes of Monas block their way, preventing attacks." (pg19)

The zone close to plant roots is called the rhizosphere. The plant releases a mix of sugars, fats and proteins into this zone, making it a rich habitat for many soil microbes, with microbe numbers generally 10–100 times greater than in the surrounding soil.

Living close to this steady supply of food, soil bacteria such as **Pseudomonas** (the 'Monas') can grow extremely fast. Here in great numbers, they help plants by physically blocking the entry of pathogenic (disease-causing) microbes, by making antimicrobials that kill or inhibit the growth of pathogenic microbes and by releasing chemicals (plant hormones) to stimulate the growth of plant roots and mycorrhizal fungi.

ACTINS: THE SOIL BUILDERS

"…the Actin bacteria make the darkness smell delicious…making the soil sticky and good to eat." (pg21)

Actinobacteria (the 'Actins') play a vital role in breaking down organic matter to create soil. Decomposition takes a long time – it can take billions of bacteria and fungi many months to completely break down a single fallen leaf. Many soil bacteria and fungi release glue-like molecules (super-sticky sugars and proteins) that help them stay fixed in one spot as they go about their work. Over time, these sticky substances create tiny soil clusters (micro-aggregates <250 μm) as other particles in the soil (e.g. minerals and microbes) are glued together. These tiny clusters can then become entangled within larger soil structures such as webs of fungal hyphae (like *Glomus*) and long chains of *Actinobacteria*, creating larger soil clusters (macro-aggregates >250 μm, often described as 'peds' or 'humus').

Soil *Actinobacteria* make an extraordinary diversity of molecules, including many of the antibiotics used in medicine. Some *Actinobacteria* create a molecule called geosmin that gives soil that special 'earthy' smell when it rains.

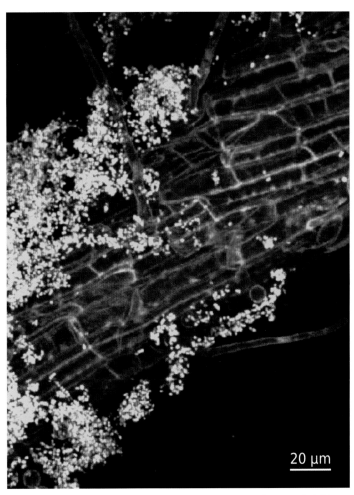

20 µm

Image of bacteria around a plant root. Sourced from Hahn L, Sá ELS, Osório Filho BD, Machado RG, Damasceno RG, Giongo A. Rhizobial Inoculation, Alone or Coinoculated with Azospirillum brasilense, Promotes Growth of Wetland Rice. Rev Bras Cienc Solo. 2016; 40: e0160006.

C. H. O. N. P.

All life forms on Earth are built from the same key **elements**: **C**arbon, **H**ydrogen, **O**xygen, **N**itrogen and **P**hosphorus, which combine to create molecules. For example:

- ○ C+H+O = SUGAR-ENERGY ($C_6H_{12}O_6$)
- C+O = CARBON DIOXIDE (CO_2)
- • H+O = WATER (H_2O)
- • O+O = OXYGEN (O_2)
- • P+O = PHOSPHATE (PO_4^{3-})
- • N+H = AMMONIUM (NH_4^+)

Most of the **energy** needed to arrange and recycle these **elements** (to grow and maintain life on Earth) comes from the Sun. Plants package this **energy** as **sugars ($C_6H_{12}O_6$)**, which fungi help transport to different members of the soil community.

This diagram shows how **energy, water** and **elements** cycle through the entire soil ecosystem.

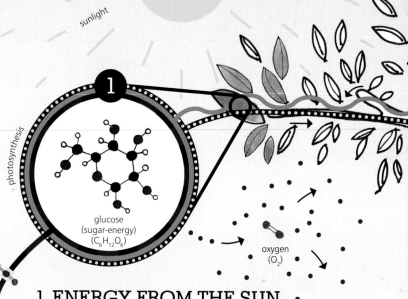

glucose (sugar-energy) ($C_6H_{12}O_6$)

photosynthesis

sunlight

oxygen (O_2)

carbon dioxide (CO_2)

1. ENERGY FROM THE SUN

Through photosynthesis, plants use **energy from sunlight** to join molecules of **water (H_2O)** and **carbon dioxide (CO_2)** to build **energy-rich sugar ($C_6H_{12}O_6$)**. This chemical reaction produces **oxygen (O_2)** as a waste product.

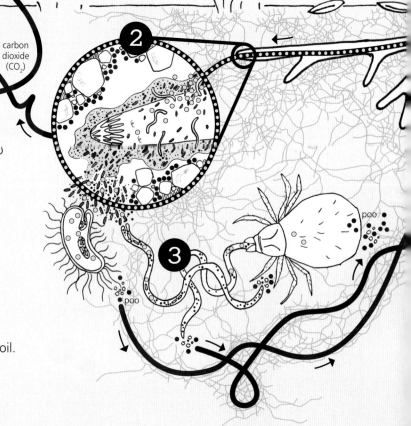

poo

poo

2. FEEDING SMALL FRIENDS

Plants release a steady supply of **energy-rich sugar** into the soil to nurture communities of microscopic bacteria, archaea and fungi around their roots. These microbes consume this **energy** and make **carbon dioxide (CO_2)** as a waste product.

3. SOIL FOOD WEBS

Smaller soil creatures (like bacteria and fungi) are eaten by larger ones, and so on. Each digests the **energy** and **elements** it needs and excretes (poos) leftover nutrients (containing **phosphorus** and **nitrogen**) back into the soil.

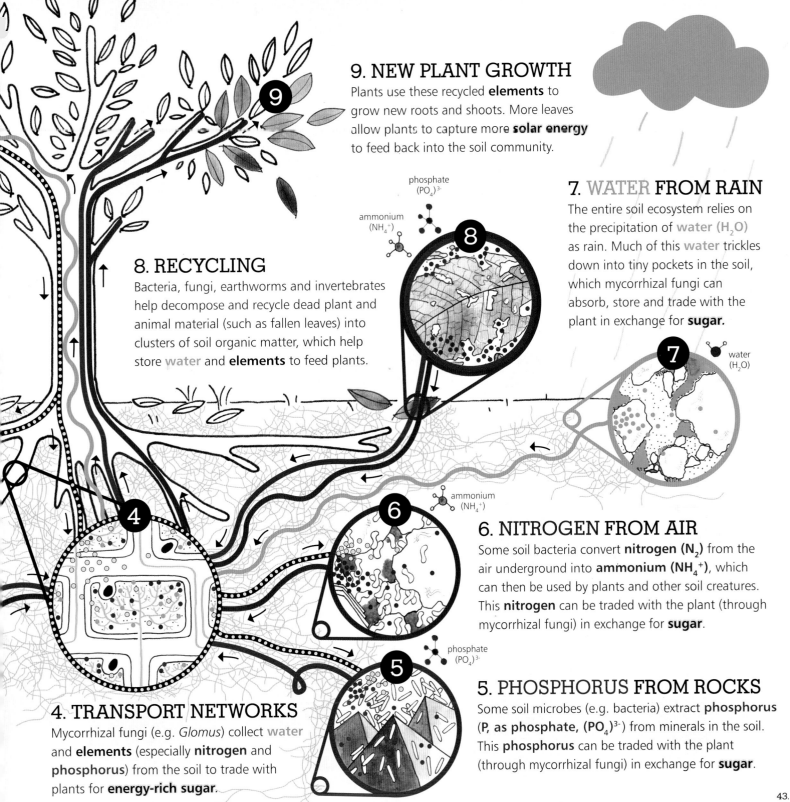

9. NEW PLANT GROWTH

Plants use these recycled **elements** to grow new roots and shoots. More leaves allow plants to capture more **solar energy** to feed back into the soil community.

7. WATER FROM RAIN

The entire soil ecosystem relies on the precipitation of water (H_2O) as rain. Much of this water trickles down into tiny pockets in the soil, which mycorrhizal fungi can absorb, store and trade with the plant in exchange for **sugar**.

phosphate (PO_4)$^{3-}$

ammonium (NH_4^+)

water (H_2O)

8. RECYCLING

Bacteria, fungi, earthworms and invertebrates help decompose and recycle dead plant and animal material (such as fallen leaves) into clusters of soil organic matter, which help store **water** and **elements** to feed plants.

ammonium (NH_4^+)

6. NITROGEN FROM AIR

Some soil bacteria convert **nitrogen** (N_2) from the air underground into **ammonium** (NH_4^+), which can then be used by plants and other soil creatures. This **nitrogen** can be traded with the plant (through mycorrhizal fungi) in exchange for **sugar**.

phosphate (PO_4)$^{3-}$

4. TRANSPORT NETWORKS

Mycorrhizal fungi (e.g. *Glomus*) collect water and **elements** (especially **nitrogen** and **phosphorus**) from the soil to trade with plants for **energy-rich sugar**.

5. PHOSPHORUS FROM ROCKS

Some soil microbes (e.g. bacteria) extract **phosphorus** (**P, as phosphate, (PO_4)$^{3-}$**) from minerals in the soil. This **phosphorus** can be traded with the plant (through mycorrhizal fungi) in exchange for **sugar**.

43.

HUMANS AND SOIL

Soil is the 'skin of the Earth'. Soil ecosystems (the pedosphere) develop through constant dynamic interactions between Earth's life forms (biosphere), rocks and minerals (lithosphere), gases (atmosphere) and water (hydrosphere).

Humans have always depended on soil. It is the source of our food (grains, fruits and vegetables), fuel (firewood), fibre (clothes and paper) and feed for livestock (hay and corn). Over the past 60 years, human use of soil for agricultural production has intensified and now covers about 40% of land on Earth. At the same time, we have seen a growth of soil-related problems, such as the loss of top soil, increases in soil erosion, soil salinity (high salt content), soil acidity (low pH) and an increase in pests. The poor health of our soils is a threat to the future stability and sustainability of human civilisation.

*

HEALTHY SOIL IS GOOD FOR THE CLIMATE

The Earth's soils hold about three times as much carbon as the atmosphere. The more soil bacteria, fungi, nematodes, protozoa, springtails and earthworms there are – dead or alive – the more carbon is stored in their bodies and surrounding soil. Activities that increase biodiversity within the soil food web remove more carbon dioxide from the atmosphere.

RETHINKING AGRICULTURE

A growing range of approaches to farming focus on working with ecological systems to enrich the biodiversity living both aboveground and belowground. Broadly described as 'regenerative agriculture', these methods include agroecology, permaculture, and biodynamic and organic farming. They **encourage** farmers and gardeners to use techniques such as composting, mulching and planting a mixture of crops, and **discourage** the use of tilling (ploughing), synthetic pesticides and synthetic fertilisers.

Synthetic pesticides are chemicals designed to kill or block creatures from causing damage to crops. However, these chemicals often have unintended consequences. For example, many pesticides kill beneficial insects like honeybees and ladybird beetles. Regenerative agriculture tries instead to work with friendly soil microbes living around plants to protect them from insect pests and disease-causing fungi or bacteria. For example, instead of using insecticides to control beetle larvae from damaging the roots of many plants, *Steinernema* nematodes (featured in the picture book *Nema and the Xenos*) can be used to stop insect pests from damaging many crops (this process is also described as biological control).

Scientists have found that the overuse of **synthetic fertilisers** can cause plants to be selfish. When they have all the nutrients they need, plants share less energy-rich sugars and fats through their roots. This gradually causes a loss of soil biodiversity and breaks their relationships with mycorrhizal fungi. Instead of synthetic fertilisers, regenerative approaches nurture soil microbes to recycle and replace nutrients taken from the soil by previous crops.

* 'Slash and burn agriculture in the Amazon' by Matt Zimmerman, (https://flic.kr/p/3jFShG). Licence at www.creativecommons.org/licenses/by/2.0

CARING FOR SOIL

We can all do our bit to keep soil happy. Healthy soil helps take carbon from the atmosphere and stores it underground, making homes for microbes. Nurturing the smallest patch of soil can make a difference.

PLANTING

Soil is healthiest when it has plants growing in it. Plants transform the Sun's energy into carbon-rich sugars and feed them to the billions of creatures living underground. The more plants there are, the more energy and carbon they can share through their roots to nurture the entire soil ecosystem.

MULCHING

Soil likes having a protective blanket. Adding a layer of leaves, straw or compost (called mulch) to soil helps stop the loss of moisture (evaporation) in summer, hold heat during winter, and block the growth of weeds around plants.

WATERING

Soil is happiest with plenty of water. However, rainfall is not reliable, and plants depend on a steady supply of water – which is why we often need to water plants. Nurturing a rich biodiversity of soil creatures also helps store water in microscopic pockets, pools and pathways in the soil.

ENCOURAGING FUNGI

Underground networks of fungi store and transport water and nutrients throughout the soil. Ploughing and tilling break up these delicate fungal networks, so avoid mixing or disturbing soil as much as possible.

COMPOSTING: NATURE'S FERTILISER

Soil microbes and plants need a steady supply of food to build soil organic matter, and compost is their favourite meal. Making compost is easy: just collect a heap of organic matter, such as grass, leaves, animal manure and food scraps, and wait a few weeks/months for bacteria, fungi, insects, mites and earthworms to break it down.

HOW SMALL ARE THE CHARACTERS?

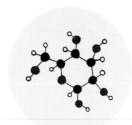

Sugar (glucose)
(700 picometres)

- Simple sugar molecule containing 6 carbon atoms, 12 hydrogen atoms and 6 oxygen atoms ($C_6H_{12}O_6$)
- Created by plants (and bacteria) in photosynthesis using energy from sunlight
- A primary source of energy in soil, traded with mycorrhizal fungi in exchange for water and nutrients

ACTINS
Streptomyces bacteria
(2 micrometres wide)

- Long chains of *Actinobacteria*
- Help decompose soil organic matter and create soil structure
- Produce huge variety of molecules to protect plants, such as antibiotics

GLOMUS SPORE
Glomus fungi
(100 micrometres)

- Reproductive part of *Glomus* fungi
- Once germinated (awoken), one or more hyphal threads extend from the spore seeking a plant root partner

---

SIZE

1,000 pm = 1 nm   1,000 nm = 1 µm   1,000 µm = 1 mm

pm
picometres (10^{-12} m)

nm
nanometres (10^{-9} m)

µm
micrometres (10^{-6} m)

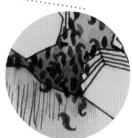

Water
(280 picometres)

- A molecule made from 2 hydrogen atoms and 1 oxygen atom (H_2O)
- Vital for all life forms
- Makes up about 20–30% of the volume of healthy soil

TILIS
Bacillus subtilis bacteria
(3 micrometres)

- Commonly found in soil, although can also live in the human gut
- Able to dissolve phosphorus (as phosphate, PO_4) from rock using enzymes

MONAS
Pseudomonas bacteria
(3 micrometres)

- Bacteria found in high numbers around plant roots
- Some are described as Plant Growth Promoting Rhizobacteria (PGPR) due to their ability to protect plants and promote plant growth

AZOES
Azospirillum bacteria
(3 micrometres)

- Commonly found in soil and around plant roots
- A type of diazotroph – able to transform nitrogen (N_2) from the air into ammonium (NH_4^+) through nitrogen fixation

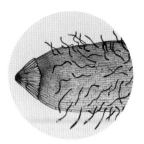

ROOT TIP

(500–1,000 micrometres)

- Growing tip of tree root
- Releases sugars and fats (called plant root exudate) to attract and feed friendly soil microbes and fungi

BROMA
Theobroma cacao

(5 metres)

- A type of tropical plant native to the Amazon basin
- Releases sugars through its roots to attract mycorrhizal fungi and soil microbes
- Beans used to make chocolate

AMAZON RAINFOREST

(3,000 kilometres wide)

- A large, terrestrial system dominated by trees
- Densely inhabited by a huge variety of plants, animals, fungi and smaller soil microbes
- The setting for this story

SUN

(1,392,684 kilometres)

- Star powered by the atomic nuclear reaction fusing hydrogen atoms into helium atoms
- Approx. 4.6 billion years old
- Provides light energy (photons) to power photosynthesis
- The most important source of energy for life on Earth

1,000 mm = 1 m   *1,000 m = 1 km*

mm
millimetres (10^{-3} m)

m
metres

km
kilometres

Sand grains

(200–2,000 micrometres)

- The largest type of soil particle
- Typically composed of silicon dioxide (SiO_2) or calcium carbonate ($CaCO_3$)

BROMA'S BABIES

(250 millimetres)

- Young *Theobroma cacao* plants
- Connection to *Glomus* mycorrhizal fungi helps them access water and mineral nutrients from soil

GLOMUS NETWORK
Glomus fungi

(1,000s of kilometres)

- System of fungal hyphae (mycelium) connecting trees within the forest system
- Also referred to as the wood wide web
- *Glomus* name derives from Latin for 'little ball of yarn'

GLOSSARY

AMMONIUM
Ammonium is a small molecule made from four hydrogen atoms joined to a nitrogen atom. Ammonium is an important source of nitrogen for many soil creatures, particularly plants.

BACTERIA
Bacteria (singular = bacterium) are the smallest single-celled life forms, usually about 1 or 2 micrometres long. Scientists have classified thousands of different species of bacteria, but it is thought there could be millions. Unlike plants and animals, bacteria reproduce by simple cell division (asexually), meaning one cell divides to make two cells, and so on. Under perfect nutrient conditions, some bacteria can reproduce every 20 minutes! Bacteria can live anywhere there is water. Their ability to consume and recycle organic matter and chemical compounds makes them an important component of all ecosystems.

CLAY
Soil minerals are usually divided into three types of particles, based on their size: sand, silt and clay. Clay forms thin, flat, hexagonal sheets (only about 1–2 μm thick), creating a large surface area for water, organic matter, minerals and microbes to attach and for chemical and biological reactions to occur. As a matrix of negatively charged aluminium, iron and silicon oxides, clay is great at trapping positively charged calcium, magnesium, potassium and ammonium, making it a great source of minerals for soil microbes and plants.

FUNGI
The word fungus (plural fungi) derives from the Greek word *sphongos* (meaning sponge). Fungi are classified into their own kingdom (*Mycota*), alongside plants and animals. In soil, most fungi release digestive enzymes (active proteins) to break down organic matter. Some fungi, such as yeast, are invisible to humans because their cells are too small (microscopic). Many live their lives underground, although some fungal structures, such as mushrooms, become visible when fruiting above ground. The gargantuan fungus (*Armillaria ostoyae*) currently holds the record as the world's largest known organism, at 8.8 km^2.

HUMUS
A commonly used (but poorly understood) term referring to small, dark, soil clusters containing communities of microbes living on minerals and other soil organic matter.

HYPHAE
Hyphae are long, thin, thread-like, tubular chains of cells, common to most fungi and some bacteria. Like plant roots, fungal hyphae grow from their tips. Fungal hyphae can extend for many kilometres through the tiny pores within soil, creating a branching network called a mycelium, which can be thought of as the fungal 'body'.

MICROBE
Short for microorganism, the term microbe describes all kinds of microscopic creatures, such as bacteria, archaea, viruses and protozoa.

MOLECULE

A molecule is a group of two or more atoms held together by chemical bonds. Some molecules, such as oxygen (O_2), water (H_2O), carbon dioxide (CO_2) and glucose ($C_6H_{12}O_6$), are simple and small. Even structures as complex as DNA and proteins can also be described as types of macromolecules.

MESSAGE MOLECULES

Plants, animals, fungi and microbes release a wide range of molecules to send signals or messages to other creatures – it's how they communicate! These message molecules (or infochemicals) can act in many different ways, to attract, to warn, or to stimulate/block the growth of other creatures. The message molecules in the story are called strigolactones, and are released by a wide range of plants to attract mycorrhizal fungi. The molecules released by the mycorrhizal fungi (called Myc factors) are also a type of message molecule.

MYCORRHIZAE

The term mycorrhizae describes the mutually beneficial symbiotic partnership between a fungus and a plant root. There are two main groups of mycorrhizae, each with different strategies to connect to plant roots. The hyphae of Ectomycorrhizae (*ektos* means outside) form a net of hyphae around the root tips of nearly 10% of all plant species, including many common forest trees (such as birch, oak, pine, fir and eucalyptus). The hyphae of Endomycorrhizae (*endo* means inside) enter individual plant root cells. About 80–85% of all plant species on Earth (including *Theobroma cacao*, the tree in this story) partner with the group of Endomycorrhizae called Arbuscular Mycorrhizal Fungi (AMF). The physical structure (morphology) of AMF has not changed for more than 400 million years and can be considered as living fossils.

NITROGEN

Nitrogen (N) is one of the main elements making up proteins and nucleic acids – essential molecular building blocks of all living creatures (along with oxygen, carbon and hydrogen). On Earth, most of the nitrogen cycles between the atmosphere (dinitrogen N_2 makes up 78% of the atmosphere) and the bodies of living organisms.

PHOSPHORUS

The element phosphorus (P) is an essential part of nucleic acids (DNA/RNA) and the cell membranes of all creatures. In nature, almost all phosphorus exists as phosphate (one phosphorus atom and four oxygen atoms, PO_4). Phosphorus is one of the most important nutrients that plants need to source from the soil, usually with the help of mycorrhizal fungi.

SPORE

Spores are the reproductive units of fungi. Fungal spores come in a wide variety of sizes, shapes and colours. They are often adapted to survive for long periods of time, particularly because many spores travel over long distances before germinating. Some fungi have evolved above-ground structures, such as mushrooms, to help spread their spores more widely through the wind and rain. Other fungi, such as truffles, attract animals to feed on them, with their spores arriving in a new home in a fresh deposit of poo. The spores of the *Glomus* fungi (in this story) can be moved by wind; however, it is thought that most are spread by small insects and animals.

MEET THE TEAM

BRIONY BARR
Art Director & Co-Director
of Scale Free Network
Briony draws on her skills as
a conceptual artist to visualise
complex systems and invisible
worlds.

DR. GREGORY CROCETTI
Science Director & Co-Director
of Scale Free Network
Gregory combines his
microbial ecology experience
with science education skills
to teach the world that
microbes are marvellous.

AVIVA REED
Illustrator
Aviva is a multidisciplinary
artist and visual ecologist
who explores complex
science through painting
and immersive installation

AILSA WILD
Writer
Ailsa creates stories for theatre
and paper pages. She loves
collaborating with acrobats,
scientists and children and her
favourite question is: But why?

Photos by Matto Lucas.

SMALL FRIENDS BOOKS

This series tells stories about symbiotic partnerships between
microbes and larger forms of life.

Each story is developed collaboratively by a core creative
team, with support and feedback from scientists, teachers
and students.

The series is co-published by CSIRO Publishing and
Scale Free Network. To learn more, visit:
www.publish.csiro.au/books/series/81

Also available in the series

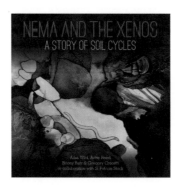